SOIL ANALYSIS

Study of Physico-Chemical Properties and Micronutrients in Soil of Wardha Region of India

By

Dr. Pradip V. Tekade

M.Sc., Ph.D., B.Ed., NET, SET, GATE

PREFACE

I am extremely happy in presenting this book , which consists of study of physico-chemical properties and micronutrients in soil of Wardha region of India. Attempts have been made to make the book free of errors and omissions. However, critical comments and concrete suggestions will be warmly and thankfully received which will helpful to improve the contents of this book in future.

During this research project and in the preparation of this book, I have referred to available books/research papers written by eminent authorities in the field, therefore I express my indebtedness to such authorities. It is my duty to express the sincere thanks to team of **Create Space**, the leading publisher for their help in the publication of this book.

I am extremely indebted to Mr. M.V. Paranjape, Mr. B. R. Pol, District soil survey and soil analysis laboratory, Wardha for extending their help and providing proper guidance.

I am thankful to University Grant Commission (WRO) for providing the financial assistance. I wish to express my sincere thanks to Principal Dr. Om Mahodaya, teachers and students of Jankidevi Bajaj College of science,

Wardha for their valuable suggestions and inspiring encouragement given to me. I appreciate the work of Mr. Arvind Tandale for his untiring help in the typesetting of this book.

It is pleasure to thank my parents, wife Mrs. Rohini, kid Bhooshit and all family members whose patience, dedication and encouragement made this work possible. I will be failing in my duties if I do not express my thanks to my well-wishers and friends who help me directly or indirectly during the tenure of this work.

Author

CONTENTS

S.N.	CHAPTER	PAGE NO.
I	INTRODUCTION	7-10
II	SIGNIFICANCE AND OBJECTIVES OF THE STUDY	11-18
III	EXPERIMENTAL STUDY	19-45
IV	SIGNIFICANCE OF FERTIILIZERS IN CROP PRODUCTION	46-51
V	RESULTS AND DISCUSSION	52-76
VI	SUMMARY AND CONCLUSIONS	77-79
	REFERENCES	80-85
	ANNEXURE	86

| CHAPTER I | INTRODUCTION |

The soil is natural medium for plant growth. Soil supplies nutrients for plant growth, and plants manufacture feed for animals and man also. Some soils are naturally productive, some are unproductive. While majority of soils, which must be fertilized, irrigated drained or limed to make them desirably productive. The productive soil is one which contains adequate amounts of all essential elements in forms readily available to plants.

In United State of America, vast study had been done on soil analysis. The work on the soil classification is also done in Russia, Newzealand and Pakistan also. In India, vast study have been done on soil and water conservation, study of colloidal and biological properties of soil, fertilizers, their uses and soil fertility etc.

Plant Nutrition

The elements required for plant growth and reproduction are Carbon, Hydrogen, Oxygen, Nitrogen, Phosphorus, Potassium, Calcium, Magnesium, Iron, Sulpher, Boron, Manganese, Copper, Zink, Molybdenum and Chlorine. Among these elements essential for plants, Boron and Molybdenum are not essential for animals, however, animals, need Sodium, Cobalt and Iodine which are not required by plants.

Micronutrients

Among the nutrients considered essential for plants, curtain nutrients like Boron, Manganese, Copper, Zinc and Molybdenum are required in trace amounts, therefore called micronutrients. The micronutrients have a very high efficiency value i.e. very small quantities are sufficient to produce optimum effects. However, slight deficiency or slight excess can cause serve damage to plants. The range from deficiency condition to that of excess is extremely short in the case of these nutrients, thus great care needs to exercise in the use of micronutrients.

Physiological Role of Essential Nutrients

1. Nitrogen:

Nitrogen is present in protoplasm in the form of proteins. In addition to this it is present in chlorophyll, nucleotides, nucleosides, phosphatides and alkaloids which are of great physical importance in metabolism.

2. Phosphorus:

It occupies key position in metabolism. It is essential constituent of many vital compounds like nucleotides and most enzymes.

3. Potassium:

   a.   Potassium increases the efficiency of the leap in manufacturing of sugars and starch.

   b.   It increases the plant's resistance to disease.

   c.   Potassium acts as activator for several enzymes involved in the protein synthesis and carbohydrate and nucleic acid metabolism.

4. *Calcium :*

   a.   Calcium helps in increasing stiffness of straw.

   b.   It acts as detoxifying agent in the neutralization of organic acids by providing the basic material.

   c.   It encourages seed production.

5. *Magnesium:*

   a.   Magnesium present in chlorophyll. The green color of plants is due to Magnesium.

   b.   Magnesium plays an essential role in many enzymatic reactions.

6. *Sulpher:*

   a.   Sulpher is constituent of enzymes proteins and coenzymes.

   b.   Sulpher stimulates root growth, seed formation and nodule formation.

Micronutrients

Micronutrients (Iron, Boron, Manganese, Copper, Zink, Molybdenum and Chlorine) used by plants in small amounts.

These elements may limit plant growth because there may not be sufficient amount of them in the soil.

Iron:

It is involved in several oxidation-reduction reactions in plants and thus essential for the synthesis of proteins and several metabolic reactions.

Manganese:

It acts as a catalyst in oxidation-reduction reactions in plants. It is also acts as an activator of many enzymes.

Zinc:

Zinc is a essential for number of enzymatic reactions. It also helps in the formation of growth hormones.

Copper:

It acts as an electron carrier in enzymes which bring about oxidation-reduction, and regulates respiratory activity in plants.

Boron:

Boron is involved in the uptake of calcium and its efficient use by plants also helps in protein synthesis.

Molybdenum:

Molybdenum is essential for nitrogen fixing organisms (symbiotic and nonsymbiotic)

Chlorine:

Chlorine is thought to be essential for photosynthetic process.

CHAPTER II	SIGNIFICANCE AND OBJECTIVES OF THE STUDY AND

Origin of the problem:

For increasing the soil fertility and crop yields the vast studies have been done India. As a result of these studies, suitable analytical methods have been recommended and fertilizer recommendation prescribed.

In order that fertilizer recommendations become more useful, more calibration studies of soil test with responses of different crops should be undertaken under all agro-climatic conditions in country. Accurate fertilizer recommendation based on soil-test-crop-response correlation studies in countries will make the soil testing service more useful and popular and will result in the more efficient use of fertilizers to obtain crop yields.

Problem: The district Wardha is situated, in the centre of India. Wardha is surrounded by small towns and villages. The main income source of the people is agriculture. The main crops yielded in this region are cotton, jowar, soyabean, sugarcane, bananas, pulses and some common vegetables.

To increase the crop growth and yield of crops, the farmers used common fertilizers containing N, K and P elements and superphosphates etc. But, for the growth of plants as mentioned earlier, micronutrients are also essential. So, it is topic of interest to test the soil and to study the physical parameters as well as to test the available nutrients present in the soil, taken from various parts of this region. After studying all these properties, the suitable fertilizer recommendation should be prescribed.

Significance of the study:

By testing the soil and studying the physical parameters as well as testing the available nutrients present in the soil taken from various parts of the Wardha district region, the suitable fertilizer recommendation should be prescribed, by which the quality of crop can be improved.

Objectives

1. To study the physical parameters and available nutrients as well as micronutrients present in the soil.

2. To suggest suitable fertilizer for the particular soil, so as to improve the quality of the soil and thus crops productivity can also be increased

The physico-chemical parameters of the soil and their significance:

Determination of soil pH:-

A pH value is a measure of Hydrogen ion activity of the soil water system and expresses the acidity and alkalinity of the soil. It is a very important property of soil as it determines the availability of nutrients, microbial activity and physical condition of soil. The pH of a solution has been defined as the negative logarithm of hydrogen ion activity, which in very dilute solution can be expressed as concentration, in gram mole / litre.

$$pH = -\log[H^+]$$

Based on the pH value the following ratings may be adopted

Table 1: pH value and characteristic of soil

S.N.	Rating	pH range	Characteristics
1.	Extremely alkaline	< 9.0	Characteristic of highly alkaline soil requiring reclamation measures
2.	Strongly alkaline	8.4 - 9.0	

3.	Moderately alkaline	7.6 - 8.3	Suitable for many of agriculture crops and pH beyond 8.0 – 8.3 can be tolerated by crop like rice, Lucerne
4.	Slightly alkaline	7.1 – 7.6	
5.	Nearly neutral	6.5 – 7.6	
6.	Slightly acidic	6.0 – 6.5	Characteristic of many red and lateritic soils indicating deficiency of Ca and low in base saturation
7.	Moderately acidic	5.3 – 6.0	
8.	Strongly acidic	4.5 – 5.2	Characteristics of soils of the humid regions where annual rainfall exceeds 80 inches and of true litreates of the Malabar coast.
9.	Extremely acidic	> 4.5	

Determination of electrical conductivity of soil:-

Specific electrical conductivity of a solution is directly related to its ion contents. It is measured with conductivity cell and expressed in milliohms or mill Siemens (mS) per cm or dS m^{-1} (desi-siemens per meter).

Determination of organic carbon content of soil:-

The organic matters play an important role in supplying water to the plants and also by providing good physical conditions to the soil system.

This method estimates the organic carbon of the soil to the extent of 77 % of the total quantity of organic carbon through oxidation and hence, to get the total amount of organic carbon the percent organic carbon is multiplied by 1.3 factors (correlation in factor 100 / 77). The soil organic matter contains about 48 to 58 % of organic matter present in soil the value is multiplied by Van Bemjmelen factor i.e., 1.724.

Recently Broadbent (1953) suggested 1.9 and 2.5 as multiplying factor for surface and sub-surface soil respectively to convert percentage of organic carbon to organic matters.

Determination of available Phosphorous in the soil:-
Phosphorous and the soil occurs as orthophosphate in several forms and combination and only a small fraction of the total amount present may be available to plants, which is of direct relevance in assessing the Phosphorous fertility level.

Determination of available Potassium in the soil:-

The total potassium contents of the soil vary from 0.05 to 2.5 %. Only a small fraction of total K is held in exchangeable form ,while the rest remain in fixed or non- exchangeable form. When the crops exhaust the supply of exchangeable K, more K is released from the fixed reserve. On the contrary, if K is added to the soil, a major part of it enters into non-exchangeable form. Exchangeable K thus tends to remain at a constant level in the soil and therefore is very useful determination. Good correlation between yields and exchangeable K has been obtained. Exchangeable K is therefore, also referred to as "available K."

Determination of available micronutrients:-

Available metallic ions (Zn, Cu, Fe, Mn)
The method commonly used for determining the available micronutrients in soil sample is given by Lindsay and Norvell (1978).

The method consists of use of DTPA (Diethylene triamine pentacetic acid) as an extractant which has been widely accepted for simultaneous extraction of micronutrient cation viz. Zinc (Zn),Copper(Cu),Iron(Fe) and Manganese(Mn) in neutral and alkaline soils. The content of these cations in the extract is determined on an Atomic Absorption Spectrophotometer (AAS).

Principle

DTPA as a chelating agent combine with free metal ions in the solution to form soluble complexes. Stability constants for the simultaneous complexing of Zn, Cu, Mn and Fe shows DTPA as a most suitable extractant. Excessive dissolution of $CaCO_3$ which may release occluded micronutrients that are not available for plants particularly in calcareous soils misleads the results. To avoid this, the extractant is buffered in slightly alkaline pH range and included soluble Ca^{++}.

Triethanolamine (TEA) is used as a buffer because it burns clearly during atomization. At the selected pH of 7.3,three fourth of TEA is protonated ($HTEA^+$) which exchanges with Ca^{++} and some Mg^{++} from the soil exchange sites.

This increases the concentration of calcium ions (Ca^{++}) by two or three folds and help in suppressing the dissolution of $CaCO_3$ in calcareous soils.

The DTPA has capacity to complex each of the micronutrient cation as 10 times of its atomic weight. The capacity ranges from 550-650 mg kg^{-1} depending on the micronutrient cation.

| CHAPTER III | EXPERIMENTAL STUDY |

Soil Sampling

The importance of having a representative samples can be well realized because of only a very small fraction of soil mass of the field is used for analysis. While collecting samples, the following points were taken into consideration.

1. Where surface samples are collected for advisory work, it must represent the area.

2. A field can be treated as a single sampling unit only if it is appreciably uniform.

3. The number of samples to be collected from a field would depend upon the variability of the land.

4. Variation in slope, cooler, texture, crop growth and management should be taken into account and separate sets of composite samples need to be collected from each of such area.

5. Recently fertilized plots, bunds, channels, marshy spots and areas near trees or compost pits or other institute non-representative locations must be avoided during sampling.

6. If the field is leveled and soil appears to be uniform type, one sample, if taken properly, could be enough for an area of about 2 to 3 hectares.

7. If farm has undulating topography bottomlands, uplands and slopes, separate sample should be drawn from each unit.

8. The root penetration habit of plants is to be considered for ascertaining the depth of sampling.

9. For the field crops, a sampling depth of 15 to 20 cm is desired.

10. For pasture crop a 10 cm depth is normally sufficient, For deep-rooted crops and for horticulture crops, samples from different depth or layers depending upon the rooting depth of the plants may be needed.

11. For estimation of immobile nutrients like P, Kand Ca, soil samples to tillage depth can give satisfactory results; however for analysis of mobile nutrients like NO_3 and SO_4, soil sample should be taken to a depth of 60 cm.

12. Samples should not be collected from area, which is within 50 cm from the border of the field.

Samples were collected from the fields, by noting the variations in slope, color, texture, management and cropping pattern. The field was then demarcated into uniform portions, each of which was samples separately. A zigzag pattern of sampling was followed, to obtain a composite sample. A representative composite soil sample was composed of 8-20 sub-samples from piece of uniform field. The soil so collected was thoroughly mixed (by hand) on a clean piece of cloth and about 500 g of composite sample was taken for analysis. After that the soil sample was dried at room temperature in shade, it was collected and stored in bags. The bags containing soil samples were labeled with respect to

1. Field Number/ Survey No.
2. Date of sampling
3. Depth of sampling and
4. Name of the cultivator along with complete postal address

Sampling tools

Following sampling tools were used satisfactorily for the soil sampling purpose.

1. Soil tube

2. Screw type auger

3. Post hole auger

4. Spade and

5. Sickle

For sampling soft and moist soil, the tube auger, spade or khurpi was found to be satisfactory. A screw auger is convenient on hard/ dry soil, while post hole auger is useful for sampling in excessive wet areas/rice fields. Where spade and Khurpi was used, a 'V' shaped hole was dug up to the plough layer and uniform thick slice of soil mass

was taken. Augers were used for sampling from lower depths.

Following precautions were taken during soil sampling.

1. Sampling from farm lanes and field borders was avoided.

2. Sampling from fertilizer bands in row of crops was avoided.

3. Sampling from any area which is distinctly different from the dominant soil type in the field such as eroded spots, small saline areas, and sandy ridges was avoided while these areas were sampled separately.

4. Sampling from dead furrows and dead lands was avoided.

5. Sampling from old manure piles or odd straw stack bottoms was avoided.

6. Sampling from the locations where brush piles have been burnt was also avoided.

Soil preparation

The air-dry soil was passed through 2-mm sieve for analysis. Before sieving, the soil clods were lightly crushed in wooden mortar and pestle. Plant residues, grovels and other foreign matter on the sieve was discarded. If the grovel is substantial, it was noted separately. For special determination for which it is necessary to weigh out small quantities of soil i.e. organic carbon, $CaCO_3$ and chemical analysis, and representative subsumable was grind and sieved it through 0.5 mm (32-mesh) sieve.

For micronutrients like copper, iron, manganese and zinc, a brass sieve was avoided and aluminum or plastic sieve with nylon netting was used. Soil samples were stored in cloths or polythene bags. Relevant information regarding slope, drainage, irrigation, previous cropping history, fertilizer used for previous crops and proposed crops etc. of the field was provided with each sample.

Methods of determination of physico-chemical parameters of soil:

Determination of soil pH:

pH measurement was done by measuring the potentials developed due to hydrogen ion concentration as it is proportional to hydrogen concentration of solution. pH measurement was done by using a pair of electrode called

1. Reference electrode and 2. Glass electrode

1. *Reference electrode:-*

It is also called calomel electrode. It is filled with saturated solution of KCl. The potential at this electrode is zero when placed in solution.

2. *Glass electrode:-*

It is hydrogen ion sensitive electrode and filled with pH sensitive solution. It has a bulb of permeable membrane at the tip of electrode made of glass.

A potential is proportional to hydrogen ion concentration of solution develops on this electrode, when put in the test solution.

When the both electrodes are kept immersed in solution, a difference of potential develops at two electrodes, which is measured by meter and expressed in terms of pH. Soil pH is determined in a soil water suspension ratio 1:2 or 1:2.5. Recently Schofield and Taylor (1955) have shown that a much reliable measurement of soil pH can be made in 0.01 M $CaCl_2$.

PH meter:

The pH being a potentiometer requires to be calibrated before use with buffer solutions of known pH values.

Reagent:

Standard buffer solution:-

Buffer solution required for calibration of pH meter may be of pH 4.0 and in other ranges (pH 7.0 and 9.2) of expected soil pH value. 1 tablet was dissolved in 100 ml double distilled water.

Procedure: - (Soil to water ratio 1:2)

1.   20 gm of soil was taken in 50 ml beaker

2.   40 ml of distilled water was added and the suspension was stirred intermittently with glass rod for 30 minutes.

3.   pH meter was standardize by using a buffer solution having 4.00 and 7.00 pH. Electrode is washed with jet of water.

4.   Electrode is inserted into the soil suspension and pH value is recorded.

5.   The electrode was removed from soil suspension and the adhering soil particles on the electrodes was washed with a fine jet of water.

6.   The electrode was wiped with tissue paper and muslin cloth.

Determination of Electrical conductivity of soil:-

Reagents

Standard potassium chloride solution:-

0.7456 g of dry potassium chloride (A.R. grade) dissolved in distilled water and the volume was made up to one litre. At 25 ^{0} C it gives an electrical conductivity of 1.41 m mhos/cm or dsm^{-1}. The conductivity meter is to be calibrated and cell constant determined with the help of this solution. Even if the scale is given directly in m mhos / cm it is necessary to check the instrument with this solution.

Procedure:

1.   20 g of the soil was taken in 50 ml beaker.

2.   40 ml of distilled water or deionised water was added and suspension was stirred intermittently for half an hour and was kept it for 30 minutes.

3.   Conductivity cell in suspension solution was inserted and electrical conductivity of the sample was noted.

Determination of organic carbon contents of soil

(Walkley and Black method)

Reagents

1.   *Potassium dichromate solution (1 N):*

Dissolve 49.04 g pure crystals of $K_2Cr_2O_7$ in distilled water and diluted to 1 litre

2.   *0.5 N Ferrous sulphate solution:*

139 g $FeSO_4$ (A.R. grade) was dissolved in distilled water and 15 ml conc. H_2SO_4 was added and diluted to 1 litre. Ferrous ammonium sulphate can also be used.

To prepare 0.5 N solutions of FAS, dissolve 196 g of salts in 800 ml distilled water containing 20 ml conc. H_2SO_4 and dilute to 1 litre.

3.   *Conc. H_2SO_4*

4.   *Orthophosphoric acid: 85 %*

5.   *Diphenylamine indicator:*

0.5 g diphenylamine was dissolved in a mixture of 100 ml conc. H_2SO_4 and 20 ml water.

6.   *Ferroin indicator (0.025 N):*

1.485 g orthophenathroline monohydrate and 0.695 g $FeSO_4.7H_2O$ was dissolved in water and diluted to 100 ml (during titration color of the solution changes from dull green to chocolate red). Indicator is also available as prepared solution. Addition of 3 to 4 drops of this indicator is sufficient. (Note: - Orthophosphoric acid is not required when ferroin solution is used as an indicator)

Procedure:

0.5 to 1.0 g finely ground soil sample was passed through 0.5 mm sieve without loss into 500 ml conical flask. 10 ml 1 N potassium dichromate solution was added by means of pipette followed by 20 ml conc. H_2SO_4 with measuring cylinder. The contents of the flask was shaken for 2 minutes and set aside on an asbestos sheet for exactly half an hour.At the end of the period 200 ml distilled water, 10 ml orthophosphoric acid and 1 ml diphenylamine indicator was added.

The contents were titrated with standard FAS or FeSO4 till color flashes from blue violet to brilliant green. Similarly blank titration was carried out without soil.

Determination of available phosphorous

The methods are commonly used for determining available phosphorous in the soil are : Olsen's Method (Olsen ET al., 1954) is used for neutral alkaline soil while the Bray and Kurtz (Bray and Kurtz 1945) is used for acid soil

Olsen's Method (NaHCO₃)

In this method phosphorous is extracted from the soil

using 0.5 M NaHCO$_3$ (pH 8.5) as an extractant. Sodium bicarbonate decreases the conc. Of Ca in solution by precipitating Ca as CaCO$_3$, as a result the conc. of P in solution is increases. Phosphorous extracted by NaHCO$_3$ is estimated calorimetrically by adding ammonium

molybdate and thereafter reducing the molybdenum phosphate complex in acidic medium. The intensity of blue color on reduction provides a measure for concentration of P in the extract.

Reagents:

1.   *0.5 M NaHCO₃ solution (pH 8.5) :-*

42 g NaHCO$_3$ was dissolved in water and made up to 1 litre. pH of solution was adjusted to 8.5 with 20 % NaOH solution (about 3 -4 ml of NaOH solution per litre is needed.

2.   *Reagent A: -*

12 g ammonium molybdate was dissolved in 250 ml distilled water. In another flask 0.2910 g antimony potassium tartarate was dissolved in 100 ml water.

Both these solutions was added to 1000 ml of 5 N H$_2$SO$_4$ (140 ml conc. H$_2$SO$_4$ in 1000 ml water) and were mixed thoroughly and 2000 ml volume was made .The solution was stored in corning glass bottle in a dark and cool place.

3.   *Reagents B:-*

1.056 g ascorbic acid was mixed with 200 ml of

reagent A. This does not keep for more than 24 hours at room temperature. The solution was prepared daily as required.

4.   *Standard stock P solution (100 ppm P)*

0.439 g potassium dihydrogen orthophosphate (KH_2PO_4) A. R. grade was dried in oven at 60 ^0C for one hour and cooled in desiccators and then was dissolved in a 500 ml distilled water. 25 ml 7 N H_2SO_4 was added to it and made to 1000 ml with distilled water. This gives 100 ppm P standard stock solution. By diluting it 50 times 2 ppm P solution was prepared.

Procedure

A) Preparation of standard curve

a. A series of standards were prepared by taking 0, 2, 4, 6, 8 and 10 ml of 2 ppm P solution in 25 ml volumetric flask separately, which correspond to 0, 0.16, 0.32, 0.48, 0.64 and 0.80 ppm P or $\mu g/ml$ P respectively.

b. 5 ml extracting solution (0.5 $NaHCO_3$) was added to each flask and the pH was adjusted as above.

c.   The contents were diluted to about 20 ml with distilled water and then 4.0 ml reagent B was added.

d. The volume was making up and the intensity of blue color at 730-840 nm was measured on Colorimeter by using red filter (660 nm).

Table 2: Determination of available phosphorous

Volume *	Final Volume (ml)	Phosphorus concentration		(P)	Colorimeter reading
		µg P in 25 ml	PPM P	µg/ ml	
0	25	0	0	0	0
1	25	2	0.08	0.08	8
2	25	4	0.16	0.16	15
4	25	8	0.32	0.32	30
6	25	12	0.48	0.48	44
8	25	16	0.64	0.64	61
10	25	20	0.80	0.80	90

*Volume of 2 ppm P (2 µg P/ml) taken   (ml)

$$\text{Factor (F)} = \frac{\text{Concentration of P}}{\text{Corresponding reading of above concentration}} = \frac{0.32}{30} = 0.01$$

Factor (F) = 1 colorimeter reading = 0.01 µg P/ml

B) Extraction of soil P

1.   2.5 g air dry soil (2mm diameter) was taken into a 250 ml conical flask.

2.   Little activated carbon was added to it.

3.   After that 50 ml Olsen's reagent (0.5 M $NaHCO_3$, pH 8.5) was added to maintain soil to solution ratio of 1: 20 and then solution was shaken on the shaking machine for 30 minutes.

4.   Similarly, blank experiment was performed without taking soil.

5.   Then solution was filtered through Whatmann No.40 filter paper into a clean and dry beaker.

C) Estimation

1.   5 ml of above prepared extract ($NaHCO_3$ / filtrate) was pipette out into 25 ml volumetric flask.

2.   1ml of 5 N H_2SO_4, two drops of 2, 4 paranitrophenol was added and again 5 N H_2SO_4 was added drop by drop with intermittent shaking till yellow color disappears.

3.   Contents were diluted to about 20 ml with distilled water andthen 4 ml reagent B was added.

4.   Then the volume was made up and intensity of blue color was measured at 600 nm using red filter on colorimeter.

Observations and Calculations

Weight of soil taken = X gm

$$P \text{ (ppm in soil)} = \frac{\text{Total volume of extract}}{\text{aliquot taken (ml)}} X \frac{1}{\text{wt .of soil (g)}}$$

P (Kg/ha) = ppm P in soil X 2.24

P_2O_5 (Kg/ha) = P (Kg/ha) X2.29

Conversion factor = P (Kg/ha) X 2.29 = P_2O_5

P_2O_5 X 0.437 =P

Determination of available Potassium

(Neutral N ammonium acetate method)

The available potassium includes the both i.e. exchangeable and water soluble forms of K present in the soil. Neutral N ammonium acetate extracts both these forms from the soil.

The NH_4 ion provides a sharp and rapid extraction of K from complex. The potassium in extract is estimated flame photometrically.

Reagents

1. Neutral N ammonium acetate solution: 77.1g reagent grade NH₄OAc was dissolved in 900 ml distilled water, pH was adjusted to 7.0 with 3N acetic acid or 3N NaOH and solution was diluted to 1 litre.

2. Standard Potassium Solution (1000 ppm K): 1.098 g oven dry AR grade crystals of KCl was dissolved in 1 litre of extracting solution.

Procedure

1. 5 gm soil was taken into 250 ml conical flask and25 ml extracting solution was added to it.

2. The content was shaken for 5 minutes on mechanical shaker.

3. Content was filtered immediately through ordinary filter paper and filtrate was collected in a beaker.

4. 5 ml of filtrate was diluted to 25 ml with distilled water.

5. The above diluted extract was atomized to flame photometer and reading was noted.

Calculations

Available K (Kg/ha) =

(RXF) x vol. of extract x DF x $\dfrac{2.24 \text{ x } 106}{\text{Soil wt.(g) x } 106}$

Available K_2O (Kg/ha) = Available K (Kg/ha) x 1.20

Where R = reading, F=Conc. of K/corresponding reading and DF = dilution factor, if any.

Determination of available nitrogen in soil

(Alkaline permanganate method)

Reagents

1. *Potassium permanganate (KMnO4) 0.32%:*

    3.2 g of $KMnO_4$ was dissolved in 1 litre of distilled water with intermittent shaking till it was completely dissolved.

The solution was stored in amber colored bottle and placed in the dark place.

2. *Sodium hydroxide solution (2.5 %):*

    25 g of NaOH pellets were dissolved in 1 litre of distilled water.

3. *Mixed indicator (Bromocresol green+ methyl Red):*

    0.01 g bromocresol green and 0.07 g methyl red were dissolved in 100 ml ethyl alcohol.

4. *Boric acid indicator solution:*

    20 gm of pure boric acid (H_3BO_3) was dissolved in about 700 ml of hot water. The cooled solution was then transferred to a 1 litre volumetric flask containing 200 ml ethanol and 20 ml mixed indicator solution.

    After mixing the contents of the flask, approximately 0.05 N NaOH was added until the color is reddish purple.

    After that 1 litre volume was made up with distilled water by mixing thoroughly.

5. *Standard sulphuric acid solution (0.02 N):*

    Working solution of sulphuric acid was prepared from 0.1 N standard H_2SO_4 solutions.

6. *Liquid paraffin*

7. *Glass beads*

Procedure

1. 20 g processed soil sample was taken into 1000 ml round bottom distillation flask.

2. 20 ml of distilled water was added to it.

3. 2 to 3 glass beads were added to prevent bumping and 1ml paraffin was also added to prevent frothing.

4. Then 100 ml of 0.32% Potassium permanganate and 100 ml of 2.5 % NaOH solution was added into distillation flask and immediately connected to Kjeldahl assembly.

5. The contents in the Kjeldahl assembly was distilled at steady rate and liberated ammonia was collected in the form of distillate in a 250 ml beaker containing 20 ml boric acid solution (with mixed indicator) during the absorption of ammonia ,pink color of boric acid solution turns green.

6. Approximately 150 ml of distillate was collected.

7. The distillate was titrated with standard H_2SO_4 solution till the color changes from green to pink.

8. Similarly blank titration was carried out without soil.

Observations:

1. Weight of soil sample taken = w g.

2. Volume of standard acid required for soil = A ml

3. Volume of standard acid required for blank = B ml

4. Normality of sulphuric acid = N

Calculations

$$\% \text{ Available N} = (A - B)X (N \text{ of acid})X0.014 \text{ X } \frac{100}{\text{wt .of soil (g)}}$$

$$\text{Available N(Kg/ha)} = \%N \text{ X } \frac{2240000}{\text{wt .of soil (g)}}$$

OR

$$\text{Available N(Kg/ha)} = (A - B)X \text{ NX } 0.014 \text{ X } \frac{2240000}{\text{wt . of soil (g)}}$$

Determination of micronutrients

Available metallic ions (Zn, Cu, Fe, Mn)

The method commonly used for determining the available micronutrients in soil sample is given by Lindsay and Norvell (1978).

The method consists of use of DTPA (Diethylene triamine pentacetic acid) as an extractant which has been widely accepted for simultaneous extraction of micronutrient cation viz. Zinc (Zn),Copper(Cu),Iron(Fe) and Manganese(Mn) in neutral and alkaline soils. The content of these cations in the extract is determined on an Atomic Absorption Spectrophotometer (AAS).

Principle

DTPA as a chelating agent combine with free metal ions in the solution to form soluble complexes. Stability constants for the Simultaneous complexing of Zn, Cu, Mn and Fe shows DTPA as a most suitable extractant. Excessive dissolution of $CaCO_3$ which may release occluded micronutrients that are not available for plants particularly in calcareous soils misleads the results.

To avoid this, the extractant is buffered in slightly alkaline pH range and included soluble Ca^{++}.Triethanolamine (TEA) is used as a buffer because it burns clearly during atomization.

At the selected pH of 7.3,three fourth of TEA is protonated ($HTEA^+$) which exchanges with Ca^{++} and some Mg^{++} from the soil exchange sites.

This increases the concentration of calcium ions (Ca^{++}) by two or three folds and help in suppressing the dissolution of $CaCO_3$ in calcareous soils. The DTPA has capacity to complex each of the micronutrient cation as 10 times of its atomic weight. The capacity ranges from 550-650 mg kg^{-1} depending on the micronutrient cation.

Apparatus

1. Analytical balances
2. Narrow mouth polythene bottles with stoppers; 100 ml capacity
3. Pipette: 20 ml capacity
4. Reciprocating electric shaker
5. Whatmann No. 1 or 42 filter papers
6. Polythene/glass funnels

7. Polythene vials, 50-ml capacity

8. Atomic Absorption Spectrophotometer (AAS).

9. Hollow cathode lamps of Zn, Cu, Fe, Mn

Reagents

1. Extracting solutions

0.005 M DTPA, 0.01 M $CaCl_2.2H_2O$ and 0.1 M TEA (Triethanolamine) adjusted to pH 7.3. 1.967 g DTPA and 13.3 ml TEA was dissolved in deionized distilled water.

1.47 g $CaCl_2.2H2O$ was added to about 500 ml deionised distilled water taken in one litre volumetric flask .To this mixture DTPA –TEA mixture was added and volume was made up to about 900 ml. pH was adjusted to 7.3 using 1N HCl. Finally the volume was made to one litre and mixed thoroughly.

2. Stock standard solutions:

The standard solutions of different micronutrients cations were prepared using AR grade chemicals. 0.1 g salt was dissolved in dilute HCl (1:1) and volume was made to 1 litre with deionised water to obtain 100 go/ ml (i.e.Mg/L or pap)solution of every micronutrient cation.

The quantity of the salt dissolved, its chemical formula, and concentration of respective stock solution is given below:

Table 3: Preparation of standard stock solutions

Element	Conc. of stock solution (μg/ ml)	Salt used	Quantity (g) dissolved in one litre of solution
Zn	100	Zinc sulphate ($ZnSO_4.7H_2O$)	0.4398
Fe	100	Ferous sulphate ($FeSO_4.7H_2O$)	0.4964
Cu	100	Copper sulphate ($CuSO_4.5H_2O$)	0.3928
Mn	100	Manganese sulphate($MnSO_4.H_2O$)	0.3076

The amount of salt as given above was dissolved in small volume of water followed by shaking after adding about 5 ml of 1:5 sulphuric acid. The contents were then diluted to 1 litre with deionised or glass distilled water.

Working standard solutions

1. *Zinc*:

10 ml of stock standard solution was taken into 100 ml volumetric flask and diluted up to the mark with DTPA extracting to get a stock solution of 10 µg Zn/ml (10 ppm).0, 1,2,4,6 and 8 ml of stock solution (10 µg Zn/ml) was taken into series of 100 ml volumetric flasks and diluted each to the mark with DTPA extracting solution. In this way standard solutions having Zinc concentration 0, 0.1, 0.2, 0.4, 0.6 and 0.8 µg/ml (10 ppm) were prepared.

2. *Iron:*

0, 1,2,4,6 and 8 ml of stock solution (100 µg Fe/ml or 100 ppm Fe) were taken into series of 100 ml volumetric flasks and diluted each to the mark with DTPA extracting solution. In this way standard solutions having iron concentration 0, 1, 2, 4, 6 and 8 µg/ml (ppm) were prepared.

3. *Copper:*

0, 1,2,4,6 and 8 ml of stock solution (100 go Cu/ml or 100 pap Cu) were taken into series of 100 ml volumetric flasks and diluted each to the mark with DTPA extracting solution.

In this way Standard solutions having copper concentration 0, 1, 2, 4, 6 and 8 µg/ml (ppm) were prepared.

4. *Manganese:*

0, 1,2,4,6 and 8 ml of stock solution (100 µg Mn/ml or 100 ppm Mn) was taken into series of 100 ml volumetric flasks and diluted each to the mark with DTPA extracting solution. In this way standard solutions having manganese concentration 0, 1, 2, 4, 6 and 8 µg/ml (ppm) were prepared.

Method

(A) Extraction of Soil Samples

1. 10g air-dried and thoroughly processed soil sample was taken into 100 ml narrow mouth polythene bottle or 100 ml conical flask.

2. 20 ml DTPA extracting solution was added to it and flask was shaken on an electric shaker for exactly two hours at 25^0 C.

3. The contents was filtered through Whatmann No.1 filter paper and ensured that the filtrate is free of colloidal matter.

4. Each sample of Zn, Cu, Fe and Mn was then analyzed on atomic absorption spectrophotometer.

5. Similarly blank experiment was also carried by adopting same procedure mentioned above without soil.

(B) Analysis of extracts

The micronutrient cations (Zn, Cu, Fe, Mn) in the soil extracts, obtained by the above described procedure was determined with the use of atomic absorption spectrophotometer as described below:

1. 'Zero' on the instrument was set.

2. Standards belonging to the element to be determined were feed to the atomic absorption spectrophotometer to standardize the instrument to read absorbance and/or concentration in the samples having the given element within the standardized range.

3. Then DTPA-extract was feed and absorbance/ concentration of the element was recorded on atomic absorption spectrophotometer.

4. The above steps were repeated for every element.

5. In some cases where instrument showed sign of 'over' for few elements in a particular sample indicating thereby that sample has a concentration out of the range for which the instrument has been standardized, then further dilution of the sample was made (2-5 times) and sample was again feed to atomic absorption spectrophotometer and absorbance/concentration was recorded.

Precautions:

1. Ensure that the deionized or double glass-distilled water used is free of the micronutrient cations.

2. The apparatus (glass/Polythene) to be used for the analysis must be thoroughly washed with acidified water and then with deionized water.

3. Shaking time, DTPA concentration, pH and temperature during shaking influence the amount of Zn, Fe, Cu and Mn extracted by DTPA. The most suitable pH for extracting solution is 7.3, shaking time two hours and temperature during shaking 25 ± 1^0C. So do not forget to adjust the DTPA extracting solution to 7.3, shaking must be carried out at 25^0C and the suspension must be filtered immediately after the shaking time of 2 hours.

4. Before feeding the extracts, it should be ensured that they are not turbid otherwise; they may block the capillary of the AAS.

CHAPTER IV	SIGNIFICANCE OF FERTIILIZERS IN CROP PRODUCTION

For vigorous growth of plants, it is necessary that all the needs of plants are met with according to their requirements. The important needs of plants are suitable physical, chemical and biological conditions of the soil. Other factors of crop production are favorable air, water and temperature regimes; freedom from insects, pests and diseases; and availability of healthy seeds of the improved variety type. Any of these factors, if not present in the right proportion at the right time, can limit the plant the plant growth rather severely.

While formulating the second five year plan, the planning commission summed up the sources available in the country for increasing food production and gave the following figures for the various inputs and their expected contributions towards the same.

Table 4: Contributions of various inputs towards increasing food production

S. N.	Input	Percentage increase attributed to the source
1	Fertilizers (Organic and inorganic)	41
2	Irrigation	27
3	Improved seeds	13
4	Double cropping in irrigated lands	10
5	Land reclamation and other sources	9

Thus, it is clear from the above table that maximum increase has been attributed to fertilizers. The need for a balanced use of fertilizers is very important. Any living body needs nutrients in a proper balance. This are true both for major and micronutrients.

Hence, special care should be given to maintenance of proper amounts of nitrogen, phosphorus and potassium; and an optimum balance amongst various micronutrients should be achieved in order to prevent toxic effects or deficiency diseases.

Micronutrients and soil fertilizers

Of the various micronutrients, the deficiency of zinc is widespread in India. However, under particular conditions the deficiency of all the other micronutrient, viz., copper, iron, manganese, boron, cobalt and molybdenum have been found to limit the crop production. Fertilization of soil either in the form of soil application or as a spray application is important. All kinds of plants are affected by micronutrient deficiencies and tremendous developments in the fertility regulation are expected in the next few years attributable to application of micronutrients.

The fertilizers recommended along with their molecular formula and % is as given below.

Table 5: Fertilizers recommended, molecular formula and percentage

Fertilizer	Chemical Formula	Content (%)	Remarks
Urea	- $CO(NH_2)_2$	Total Nitrogen content- 45% (Amide)	Acid forming
Muriate of potash	KCl	Water soluble K_2O- 60%	-

Single super phosphate	Ca $(H_2PO_4)_2.H_2O$	P_2O_5 content -16%	12% sulpher
Zinc sulphate	$ZnSO_4.7H_2O$	Zn-21%	15% sulpher
Manganese sulphate	$MnSO_4.H_2O$	Mn-30.5%	17% sulpher
Ferrous sulphate	$FeSO_4.7H_2O$	Fe-19%(Ferrous)	19% Sulpher
Copper sulphate	$CuSO_4.5H_2O$	Cu-24%	13% Sulpher

In addition to interrelationship among soil, crop and fertilizer, crop rotation and management should also be considered. The fertilizer applications should be supplemented with the additions of farm manures, crop residues and green manures, especially under scarcity conditions of fertilizers and their high cost. The residual effect of fertilizers additions must also be taken into account.

In summary, it may be reiterated that fertilizer practices are an important phase of the fertility management of soils in crop production

Table 6 : Various ranges and fertilizers recommendation and doses for the crops

S. N.	Range	Carbon	Phosphorous	Potassium
1	Very less	0-0.20	0-15	0-120
2	Less	0.21-0.40	16-30	121-180
3	Medium	0.41-0.60	31-50	181-240
4	Sufficiently enough	0.61-0.80	51-65	241-300
5	Enough	0.81-1.00	66-80	301-360
6	Very high	Greater than 1	Greater than 80	Greater than 360

Table 7 : Various ranges and fertilizers recommendation and doses for the crops

S.N.	Range	Recommendation
1	Very less	Increase the standard dose by 50%
2	Less	Increase the standard dose by 25%
3	Medium	Keep the dose as it is
4	Sufficiently enough	Decrease the standard dose by 10%
5	Enough	Decrease the standard dose by 25%
6	Very high	Decrease the standard dose by 50%

Recommended Nitrogen x 2.17 = --- Kg urea to be given

Recommended Phosphorous x 6.25 = --- Kg single super phosphate to be given

Recommended Potassium x 1.67 = --- Kg muirate of potash to be given

CHAPTER V	RESULTS AND DISCUSSION

About 35 soil samples from different villages of Wardha were collected and parameters (pH, electrical conductivity, etc.) as well as available nutrients (carbon, available nitrogen, phosphorus, and potassium) present in the soil, taken from various parts of this region were detected. Available micronutrients (Zinc, Copper, Iron and Manganese) were also determined by known methods. After studying all these properties, the suitable fertilizer recommendation were prescribed so that quality of soil can be improved and thus crops productivity can also be increased.

The details information about the different parameters which were studied and recommendation of fertilizers for different crops along with the name and address of cultivators is tabulated as follows.

Table 8-42: Report of analysis of soil and fertilizer recommendation

Farmer's Name :-	Shri.Babarao Bhujade				Sr. No.	General Analysis of soil	Observation	Remark
Address:-	At.Po- Mandaogad				1	pH	8.04	General
					2	Electrical Conductivity	0.12	General
Taluka:-	Wardha				3	Organic Carbon %(C)	0.57	Medium
District:-	Wardha				4	Nitrogen (N)	399	Medium
Survey No:-	47				5	Phosphorous (P)	15.75	Less
					6	Pottassium (K)	75	Very less

Sr. No.	Crop	Compost Manure	Total quantity of fertilizers			Details	Fertilizers to be given to crops		
			Nitrogen	Phosphoro	Pottassiu		Urea	Single	Murate of
1	Cotton	7	50	25	25	At the time of sowing	54	195	63
			50	31	38	After 30 days	54		
2	Soyabeen	5	30	75		At the time of sowing	65	586	
			30	94					
3	Jowar	6	80	40	40	At the time of sowing	87	313	100
			80	50	60	After 30 days	87		
4	Graundnut	5	25	50		At the time of sowing	54	391	
			25	62.5					
5	Pigeon	5	25	50		At the time of sowing	54	391	
			25	62.5					

Sr. No.	Micronutrient	General Percentage (P.P.M.)	Observation (P.P.M.)	Recommendations
1	Copper (Cu)	0.2	2.48	Sufficient
2	Iron (Fe)	4.5	6.83	Sufficient
3	Magnese (Mn)	2	9.19	Sufficient
4	Zinc (Zn)	0.65	0.73	Sufficient

Farmer's Name :-	Shri.Bandoo Sonule				Sr. No.	General Analysis of soil	Observation	Remark
Address:-	At.Po-Dhotra				1	pH	8.1	General
					2	Electrical Conductivity	0.13	General
Taluka:-	Wardha				3	Organic Carbon %(C)	0.4	Less
District:-	Wardha				4	Nitrogen (N)	280	Less
Survey No:-6/2					5	Phosphorous (P)	13.28	Very less
					6	Pottassium (K)	42	Very less

Sr. No.	Crop	Compost Manure (Ton per Hector)	Total quantity of fertilizers			Details	Fertilizers to be given to crops		
			Nitrogen (N)	Phosphorous (P)	Pottassium (K)		Urea	Single Super Phosphate	Murate of Potash
1	Cotton	7	50	25	25	At the time of sowing	68	234	63
			65	38	38	After 30 days	68		
2	Soyabeen	5	30 / 38	75 / 113		At the time of sowing	81	703	
3	Jowar	6	80	40	40	At the time of sowing	109	375	100
			100	60	60	After 30 days	109		
4	Graundnut	5	25 / 31.25	50 / 75		At the time of sowing	68	469	
5	Pigeon	5	25 / 31.25	50 / 75		At the time of sowing	68	469	

Sr. No.	Micronutrient	General Percentage (P.P.M.)	Observation (P.P.M.)	Recommendations
1	Copper (Cu)	0.2	1.67	Sufficient
2	Iron (Fe)	4.5	6.14	Sufficient
3	Magnese (Mn)	2	8.7	Sufficient
4	Zinc (Zn)	0.65	0.49	Zinc Sulphate directly

Farmer's Name :-	Shri.Shravan Wanjari				Sr. No.	General Analysis of soil	Observation	Remark
Address:-	At.Po- Allipur				1	pH	8.04	General
					2	Electrical Conductivity	0.14	General
Taluka:-	Hinganghat				3	Organic Carbon %(C)	0.12	Very less
District:-	Wardha				4	Nitrogen (N)	84	Very less
Survey No:-					5	Phosphorous (P)	18.9	Less
					6	Pottassium (K)	141	Less

Sr. No.	Crop	Compost Manure (Ton per Hector)	Total quantity of fertilizers			Details	Fertilizers to be given to crops		
			Nitrogen (N)	Phosphorous (P)	Pottassium (K)		Urea	Single Super Phosphate	Murate of Potash
1	Cotton	7	50	25	25	At the time of sowing	81	195	52
			75	31	31	After 30 days	81		
2	Soyabeen	5	30	75		At the time of sowing	98	586	
			45	94					
3	Jowar	6	80	40	40	At the time of sowing	130	313	84
			120	50	50	After 30 days	130		
4	Graundnut	5	25	50		At the time of sowing	81	391	
			37.5	62.5					
5	Pigeon	5	25	50		At the time of sowing	81	391	
			37.5	62.5					

Sr. No.	Micronutrient	General Percentage (P.P.M.)	Observation (P.P.M.)	Recommendations
1	Copper (Cu)	0.2	2.26	Sufficient
2	Iron (Fe)	4.5	6.33	Sufficient
3	Magnese (Mn)	2	9.74	Sufficient
4	Zinc (Zn)	0.65	1.59	Sufficient

Farmer's Name :-	Shri Vinod Bhalme			Sr. No.	General Analysis of soil	Observation	Remark
Address:-	At.Po-Allipir			1	pH	8.3	General
				2	Electrical Conductivity	0.13	General
Taluka:-	Hinganghat			3	Organic Carbon %(C)	0.9	enough
District:-	Wardha			4	Nitrogen (N)	630	enough
Survey No:-				5	Phosphorous (P)	9.45	Very less
				6	Pottassium (K)	145	Less

Sr. No.	Crop	Compost Manure (Ton per Hector)	Total quantity of fertilizers			Details	Fertilizers to be given to crops		
			Nitrogen (N)	Phosphorous (P)	Pottassium (K)		Urea	Single Super Phosphate	Murate of Potash
1	Cotton	7	50	25	25	At the time of sowing	41	234	52
			38	38	31	After 30 days	41		
2	Soyabeen	5	30 / 23	75 / 113		At the time of sowing	49	703	
3	Jowar	6	80	40	40	At the time of sowing	65	375	84
			60	60	50	After 30 days	65		
4	Graundnut	5	25 / 18.75	50 / 75		At the time of sowing	41	469	
5	Pigeon	5	25 / 18.75	50 / 75		At the time of sowing	41	469	

Sr. No.	Micronutrient	General Percentage (P.P.M.)	Observation (P.P.M.)	Recommendations
1	Copper (Cu)	0.2	1.91	Sufficient
2	Iron (Fe)	4.5	5.93	Sufficient
3	Magnese (Mn)	2	9.34	Sufficient
4	Zinc (Zn)	0.65	1.33	Sufficient

Farmer's Name :-	Ganbaji Madavi			Sr. No.	General Analysis of soi	Observatio	Remark
Address:-	At.Po-Ajangaon			1	pH	8.1	General
				2	Electrical Conductivity	0.13	General
Taluka:-	Deoli			3	Organic Carbon %(C)	0.73	iciently-eno
District:-	Wardha			4	Nitrogen (N)	511	iciently-eno
Survey No:-				5	Phosphorous (P)	9	Very less
				6	Pottassium (K)	95	Very less

Sr. No.	Crop	Compost Manure (Ton per Hector)	Total quantity of fertilizers			Details	Fertilizers to be given to crops		
			Nitrogen (N)	Phosphoro us (P)	Pottassiu m (K)		Urea	Single Super Phosphate	Murate of Potash
1	Cotton		50	25	25	At the time of sowing	49	234	63
		7	45	38	38	After 30 days	49		
2	Soyabeen	5	30 / 27	75 / 113		At the time of sowing	59	703	
3	Jowar	6	80	40	40	At the time of sowing	78	375	100
			72	60	60	After 30 days	78		
4	Graundnut	5	25 / 22.5	50 / 75		At the time of sowing	49	469	
5	Pigeon	5	25 / 22.5	50 / 75		At the time of sowing	49	469	

Sr. No.	Micronutrient	General Percentag e (P.P.M.)	Observatio n (P.P.M.)	Recommendations
1	Copper (Cu)	0.2	1.55	Sufficient
2	Iron (Fe)	4.5	5.52	Sufficient
3	Magnese (Mn)	2	8.79	Sufficient
4	Zinc (Zn)	0.65	0.53	Zinc Sulphate directly d

Farmer's Name :-		Shri. Atmaram Girde			Sr. No.	General Analysis of soil	Observation	Remark
Address:-		At.Po-Ajangaon			1	pH	8.04	General
					2	Electrical Conductivity	0.14	General
Taluka:-		Deoli			3	Organic Carbon %(C)	0.14	Very less
District:-		Wardha			4	Nitrogen (N)	98	Very less
Survey No:-					5	Phosphorous (P)	15.3	less
					6	Pottassium (K)	94	Very less

Sr. No.	Crop	Compost Manure (Ton per Hector)	Total quantity of fertilizers			Details	Fertilizers to be given to crops		
			Nitrogen (N)	Phosphorous (P)	Pottassium (K)		Urea	Single Super Phosphate	Murate of Potash
1	Cotton		50	25	25	At the time of sowing	81	195	63
		7	75	31	38	After 30 days	81		
2	Soyabeen	5	30 / 45	75 / 94		At the time of sowing	98	586	
3	Jowar	6	80	40	40	At the time of sowing	130	313	100
			120	50	60	After 30 days	130		
4	Graundnut	5	25 / 37.5	50 / 62.5		At the time of sowing	81	391	
5	Pigeon	5	25 / 37.5	50 / 62.5		At the time of sowing	81	391	

Sr. No.	Micronutrient	General Percentage (P.P.M.)	Observation (P.P.M.)	Recommendations
1	Copper (Cu)	0.2	1.57	Sufficient
2	Iron (Fe)	4.5	5.36	Sufficient
3	Magnese (Mn)	2	9.71	Sufficient
4	Zinc (Zn)	0.65	1.97	Sufficient

Farmer's Name :-	Shri.Rangdeo Mahure			Sr. No.	General Analysis of soil	Observation	Remark
Address:-	At.Po- Morchapur			1	pH	7.9	General
				2	Electrical Conductivity	0.14	General
Taluka:-	Seloo			3	Organic Carbon %(C)	0.15	Very less
District:-	Wardha			4	Nitrogen (N)	105	Very less
Survey No:-	33			5	Phosphorous (P)	27.67	less
				6	Pottassium (K)	142	less

Sr. No.	Crop	Compost Manure (Ton per Hector)	Total quantity of fertilizers			Details	Fertilizers to be given to crops		
			Nitrogen (N)	Phosphorous (P)	Pottassium (K)		Urea	Single Super Phosphate	Murate of Potash
1	Cotton		50	25	25	At the time of sowing	81	195	52
		7	75	31	31	After 30 days	81		
2	Soyabeen	5	30 / 45	75 / 94		At the time of sowing	98	586	
3	Jowar	6	80	40	40	At the time of sowing	130	313	84
			120	50	50	After 30 days	130		
4	Graundnut	5	25 / 37.5	50 / 62.5		At the time of sowing	81	391	
5	Pigeon	5	25 / 37.5	50 / 62.5		At the time of sowing	81	391	

Sr. No.	Micronutrient	General Percentage (P.P.M.)	Observation (P.P.M.)	Recommendations
1	Copper (Cu)	0.2	3.6	Sufficient
2	Iron (Fe)	4.5	5.4	Sufficient
3	Magnese (Mn)	2	8.26	Sufficient
4	Zinc (Zn)	0.65	0.98	Sufficient

Farmer's Name :-	Shri. Pandurang Girde		Sr. No.	General Analysis of soil	Observation	Remark
Address:-	At.Po-Morchapur		1	pH	7.4	General
			2	Electrical Conductivity	0.13	General
Taluka:-	Seloo		3	Organic Carbon %(C)	0.1	Very less
District:-	Wardha		4	Nitrogen (N)	70	Very less
Survey No:-			5	Phosphorous (P)	21.15	less
			6	Pottassium (K)	132	less

Sr. No.	Crop	Compost Manure (Ton per Hector)	Total quantity of fertilizers			Details	Fertilizers to be given to crops		
			Nitrogen (N)	Phosphorous (P)	Pottassium (K)		Urea	Single Super Phosphate	Murate of Potash
1	Cotton	7	50	25	25	At the time of sowing	81	195	52
			75	31	31	After 30 days	81		
2	Soyabeen	5	30 / 45	75 / 94		At the time of sowing	98	586	
3	Jowar	6	80	40	40	At the time of sowing	130	313	84
			120	50	50	After 30 days	130		
4	Graundnut	5	25 / 37.5	50 / 62.5		At the time of sowing	81	391	
5	Pigeon	5	25 / 37.5	50 / 62.5		At the time of sowing	81	391	

Sr. No.	Micronutrient	General Percentage (P.P.M.)	Observation (P.P.M.)	Recommendations
1	Copper (Cu)	0.2	3.82	Sufficient
2	Iron (Fe)	4.5	5.6	Sufficient
3	Magnese (Mn)	2	6.81	Sufficient
4	Zinc (Zn)	0.65	1.01	Sufficient

Farmer's Name :-	Shri.Anil Umate			Sr. No.	General Analysis of soil	Observation	Remark
Address:-	At.Po-Morchapur			1	pH	8.2	General
				2	Electrical Conductivity	0.14	General
Taluka:-	Seloo			3	Organic Carbon %(C)	0.16	Very less
District:-	Wardha			4	Nitrogen (N)	112	Very less
Survey No:-	107			5	Phosphorous (P)	24.75	less
				6	Pottassium (K)	168	less

Sr. No.	Crop	Compost Manure (Ton per Hector)	Total quantity of fertilizers			Details	Fertilizers to be given to crops		
			Nitrogen (N)	Phosphorous (P)	Pottassium (K)		Urea	Single Super Phosphate	Murate of Potash
1	Cotton	7	50	25	25	At the time of sowing	81	195	52
			75	31	31	After 30 days	81		
2	Soyabeen	5	30 / 45	75 / 94		At the time of sowing	98	586	
3	Jowar	6	80	40	40	At the time of sowing	130	313	84
			120	50	50	After 30 days	130		
4	Graundnut	5	25 / 37.5	50 / 62.5		At the time of sowing	81	391	
5	Pigeon	5	25 / 37.5	50 / 62.5		At the time of sowing	81	391	

Sr. No.	Micronutrient	General Percentage (P.P.M.)	Observation (P.P.M.)	Recommendations
1	Copper (Cu)	0.2	3.6	Sufficient
2	Iron (Fe)	4.5	6.01	Sufficient
3	Magnese (Mn)	2	8.26	Sufficient
4	Zinc (Zn)	0.65	1.12	Sufficient

Farmer's Name :-	Shri.Babarao Urade			Sr. No.	General Analysis of soil	Observation	Remark
Address:-	At.Po- Morchapur			1	pH	8.1	General
				2	Electrical Conductivity	0.13	General
Taluka:-	Seloo			3	Organic Carbon %(C)	0.51	medium
District:-	Wardha			4	Nitrogen (N)	357	medium
Survey No:-	219			5	Phosphorous (P)	17.1	less
				6	Pottassium (K)	38	Very less

Sr. No.	Crop	Compost Manure (Ton per Hector)	Total quantity of fertilizers			Details	Fertilizers to be given to crops		
			Nitrogen (N)	Phosphorous (P)	Pottassium (K)		Urea	Single Super Phosphate	Murate of Potash
1	Cotton		50	25	25	At the time of sowing	54	195	63
		7	50	31	38	After 30 days	54		
2	Soyabeen	5	30	75		At the time of sowing	65	586	
			30	94					
3	Jowar	6	80	40	40	At the time of sowing	87	313	100
			80	50	60	After 30 days	87		
4	Graundnut	5	25	50		At the time of sowing	54	391	
			25	62.5					
5	Pigeon	5	25	50		At the time of sowing	54	391	
			25	62.5					

Sr. No.	Micronutrient	General Percentage (P.P.M.)	Observation (P.P.M.)	Recommendations
1	Copper (Cu)	0.2	3.71	Sufficient
2	Iron (Fe)	4.5	5.77	Sufficient
3	Magnese (Mn)	2	8.43	Sufficient
4	Zinc (Zn)	0.65	1.02	Sufficient

Farmer's Name :-	Vijay Kohachade			Sr. No.	General Analysis of soil	Observation	Remark
Address:-	At.Po-Goji			1	pH	8.2	General
				2	Electrical Conductivity	0.14	General
Taluka:-	Wardha			3	Organic Carbon %(C)	0.4	less
District:-	Wardha			4	Nitrogen (N)	280	less
Survey No:-				5	Phosphorous (P)	17.55	less
				6	Pottassium (K)	43	Very less

Sr. No.	Crop	Compost Manure (Ton per Hector)	Total quantity of fertilizers			Details	Fertilizers to be given to crops		
			Nitrogen (N)	Phosphorous (P)	Pottassium (K)		Urea	Single Super Phosphate	Murate of Potash
1	Cotton	7	50	25	25	At the time of sowing	68	195	63
			63	31	38	After 30 days	68		
2	Soyabeen	5	30 / 38	75 / 94		At the time of sowing	81	586	
3	Jowar	6	80	40	40	At the time of sowing	109	313	100
			100	50	60	After 30 days	109		
4	Graundnut	5	25 / 31.5	50 / 62.5		At the time of sowing	68	391	
5	Pigeon	5	25 / 31.5	50 / 62.5		At the time of sowing	68	391	

Sr. No.	Micronutrient	General Percentage (P.P.M.)	Observation (P.P.M.)	Recommendations
1	Copper (Cu)	0.2	1.66	Sufficient
2	Iron (Fe)	4.5	6.54	Sufficient
3	Magnese (Mn)	2	8.6	Sufficient
4	Zinc (Zn)	0.65	0.49	Sufficient

Farmer's Name :-	:-Shrikant Satpute			Sr. No.	General Analysis of soi	Observatio	Remark
Address:-	At.Po-Goji			1	pH	8.4	General
				2	Electrical Conductivity	0.13	General
Taluka:-	Wardha			3	Organic Carbon %(C)	0.61	iciently-enough
District:-	Wardha			4	Nitrogen (N)	427	iciently-enough
Survey No:-	344			5	Phosphorous (P)	11.25	Very less
				6	Pottassium (K)	122	less

Sr. No.	Crop	Compost Manure (Ton per Hector)	Total quantity of fertilizers			Details	Fertilizers to be given to crops		
			Nitrogen (N)	Phosphorous (P)	Pottassium (K)		Urea	Single Super Phosphate	Murate of Potash
1	Cotton		50	25	25	At the time of sowing	49	234	52
		7	45	38	31	After 30 days	49		
2	Soyabeen	5	30	75		At the time of sowing	59	703	
			27	113					
3	Jowar	6	80	40	40	At the time of sowing	78	375	84
			72	60	50	After 30 days	78		
4	Graundnut	5	25	50		At the time of sowing	49	469	
			22.5	75					
5	Pigeon	5	25	50		At the time of sowing	49	469	
			22.5	75					

Sr. No.	Micronutrient	General Percentage (P.P.M.)	Observatio n (P.P.M.)	Recommendations
1	Copper (Cu)	0.2	2.47	Sufficient
2	Iron (Fe)	4.5	6.83	Sufficient
3	Magnese (Mn)	2	9.19	Sufficient
4	Zinc (Zn)	0.65	0.73	Sufficient

Farmer's Name :-	Prashant Satpute			Sr. No.	General Analysis of soil	Observation	Remark
Address:-	Goji			1	pH	8.3	General
				2	Electrical Conductivity	0.13	General
Taluka:-	Wardha			3	Organic Carbon %(C)	0.11	Very less
District:-	Wardha			4	Nitrogen (N)	77	Very less
Survey No:-	345			5	Phosphorous (P)	18.68	less
				6	Pottassium (K)	62	Very less

Sr. No.	Crop	Compost Manure (Ton per Hector)	Total quantity of fertilizers			Details	Fertilizers to be given to crops		
			Nitrogen (N)	Phosphorous (P)	Pottassium (K)		Urea	Single Super Phosphate	Murate of Potash
1	Cotton		50	25	25	At the time of sowing	81	19	63
		7	75	31	38	After 30 days	81		
2	Soyabeen	5	30 / 45	75 / 94		At the time of sowing	98	586	
3	Jowar	6	80	40	40	At the time of sowing	130	313	100
			120	50	60	After 30 days	130		
4	Graundnut	5	25 / 37.5	50 / 62.5		At the time of sowing	81	391	
5	Pigeon	5	25 / 37.5	50 / 62.5		At the time of sowing	81	391	

Sr. No.	Micronutrient	General Percentage (P.P.M.)	Observation (P.P.M.)	Recommendations
1	Copper (Cu)	0.2	2.46	Sufficient
2	Iron (Fe)	4.5	6.82	Sufficient
3	Magnese (Mn)	2	9.19	Sufficient
4	Zinc (Zn)	0.65	0.76	Sufficient

Farmer's Name :-	Prashant Dadaji Kutarmare				Sr. No.	General Analysis of soil	Observation	Remark
Address:-	Khandala				1	pH	7.7	General
					2	Electrical Conductivity	0.16	General
Taluka:-	Samudrapur				3	Organic Carbon %(C)	0.42	Medium
District:-	Wardha				4	Nitrogen (N)	294	Medium
Survey No:-	132				5	Phosphorous (P)	20.72	Less
					6	Pottassium (K)	160	Less

Sr. No.	Crop	Compost Manure (Ton per Hector)	Total quantity of fertilizers			Details	Fertilizers to be given to crops		
			Nitrogen (N)	Phosphorous (P)	Pottassium (K)		Urea	Single Super Phosphate	Murate of Potash
1	Cotton		50	25	25	At the time of sowing	54	195	52
		7	50	31	31	After 30 days	54		
2	Soyabeen	5	30 / 30	75 / 94		At the time of sowing	65	586	
3	Jowar	6	80	40	40	At the time of sowing	87	313	84
			80	50	50	After 30 days	87		
4	Graundnut	5	25 / 25	50 / 62.5		At the time of sowing	54	391	
5	Pigeon	5	25 / 25	50 / 62.5		At the time of sowing	54	391	

Farmer's Name :-	Maroti Khadatkar				Sr. No.	General Analysis of soil	Observation	Remark
Address:-	Waygaon				1	pH	8.01	General
					2	Electrical Conductivity	0.13	General
Taluka:-	Samudrapur				3	Organic Carbon %(C)	0.28	Less
District:-	Wardha				4	Nitrogen (N)	196	Less
Survey No:-6/2	526				5	Phosphorous (P)	19.32	less
					6	Pottassium (K)	191	Medium

Sr. No.	Crop	Compost Manure (Ton per Hector)	Total quantity of fertilizers			Details	Fertilizers to be given to crops		
			Nitrogen (N)	Phosphorous (P)	Pottassium (K)		Urea	Single Super Phosphate	Murate of Potash
1	Cotton		50	25	25	At the time of sowing	61	195	42
		7	56	31	25	After 30 days	61		
2	Soyabeen	5	30 / 38	75 / 94		At the time of sowing	81	588	
3	Jowar	6	80	40	40	At the time of sowing	109	313	69
			100	50	40	After 30 days	109		
4	Graundnut	5	25 / 31.25	50 / 63		At the time of sowing	68	394	
5	Pigeon	5	25 / 31.25	50 / 63		At the time of sowing	68	394	

Farmer's Name :-	Sunil Hiwanj			Sr. No.	General Analysis of soil	Observation	Remark
Address:-	Chincholi			1	pH	8.1	General
				2	Electrical Conductivity	0.15	General
Taluka:-	Samudrapur			3	Organic Carbon %(C)	0.24	less
District:-	Wardha			4	Nitrogen (N)	168	less
Survey No:-	52			5	Phosphorous (P)	24.45	Less
				6	Pottassium (K)	430	very very high

Sr. No.	Crop	Compost Manure (Ton per Hector)	Total quantity of fertilizers			Details	Fertilizers to be given to crops		
			Nitrogen (N)	Phosphorous (P)	Pottassium (K)		Urea	Single Super Phosphate	Murate of Potash
1	Cotton		50	25	25	At the time of sowing	68.5	195	22
		7	63	31	13	After 30 days	68.5		
2	Soyabeen	5	30 / 38	75 / 94		At the time of sowing	82	586	
3	Jowar	6	80	40	40	At the time of sowing	242	375	33
			120	50	20	After 30 days	67	391	
4	Graundnut	5	25 / 31	50 / 62.5		At the time of sowing	67	391	
5	Pigeon	5	25 / 31	50 / 62.5		At the time of sowing	81	391	

Farmer's Name :-	Shriram Timande			Sr. No.	General Analysis of soil	Observation	Remark
Address:-	Waldhur			1	pH	7.7	General
				2	Electrical Conductivity	0.1	General
Taluka:-	Hinghanghat			3	Organic Carbon %(C)	0.12	Very less
District:-	Wardha			4	Nitrogen (N)	84	Very less
Survey No:-	97			5	Phosphorous (P)	18.9	Less
				6	Pottassium (K)	211	Medium

Sr. No.	Crop	Compost Manure (Ton per Hector)	Total quantity of fertilizers			Details	Fertilizers to be given to crops		
			Nitrogen (N)	Phosphorous (P)	Pottassium (K)		Urea	Single Super Phosphate	Murate of Potash
1	Cotton		50	25	25	At the time of sowing	81.5	194	42
		7	75	31	25	After 30 days	81.5		
2	Soyabeen	5	30 / 45	75 / 94		At the time of sowing	98	586	
3	Jowar	6	80	40	40	At the time of sowing	132	313	67
			120	50	40	After 30 days	132		
4	Graundnut	5	25 / 38	50 / 62.5		At the time of sowing	82	391	
5	Pigeon	5	25 / 38	50 / 62.5		At the time of sowing	82	391	

Farmer's Name :-	Sau.Suman Lokhande				Sr. No.	General Analysis of soil	Observation	Remark
Address:-	Waygaon				1	pH	7.99	General
					2	Electrical Conductivity	0.18	General
Taluka:-	Deoli				3	Organic Carbon %(C)	0.9	enough
District:-	Wardha				4	Nitrogen (N)	630	enough
Survey No:-	407				5	Phosphorous (P)	4.5	Very less
					6	Pottassium (K)	285	iciently-enough

Sr. No.	Crop	Compost Manure (Ton per Hector)	Total quantity of fertilizers			Details	Fertilizers to be given to crops		
			Nitrogen (N)	Phosphorous (P)	Pottassium (K)		Urea	Single Super Phosphate	Murate of Potash
1	Cotton		50 /	25 /	25 /	At the time of sowing	41	234	38
		7	37.5	38	22.5	After 30 days	41		
2	Soyabeen	5	30 / 22.5	75 / 113		At the time of sowing	49	703	
3	Jowar	6	80 /	40 /	40 /	At the time of sowing	65	375	60
			60	60	36	After 30 days	65		
4	Graundnut	5	25 / 19	50 / 75		At the time of sowing	41	469	
5	Pigeon	5	25 / 19	50 / 75		At the time of sowing	41	469	

Farmer's Name :-	Sitaram Kawle				Sr. No.	General Analysis of soil	Observation	Remark
Address:-	Dhanodi				1	pH	7.93	General
					2	Electrical Conductivity	0.19	General
Taluka:-	Arvi				3	Organic Carbon %(C)	0.58	Medium
District:-	Wardha				4	Nitrogen (N)	406	Medium
Survey No:-					5	Phosphorous (P)	25.5	less
					6	Pottassium (K)	306.4	Enough

Sr. No.	Crop	Compost Manure (Ton per Hector)	Total quantity of fertilizers			Details	Fertilizers to be given to crops		
			Nitrogen (N)	Phosphorous (P)	Pottassium (K)		Urea	Single Super Phosphate	Murate of Potash
1	Cotton		50 /	25 /	25 /	At the time of sowing	55	195	32
		7	50	31	19	After 30 days	55		
2	Soyabeen	5	30 / 30	75 / 94		At the time of sowing	65	586	
3	Jowar	6	80 /	40 /	40 /	At the time of sowing	87	313	50
			80	50	30	After 30 days	87		
4	Graundnut	5	25 / 25	50 / 62.5		At the time of sowing	54	391	
5	Pigeon	5	25 / 25	50 / 62.5		At the time of sowing	54	391	

3+6/8		Pundlik Chaudhari			Sr. No.	General Analysis of soil	Observation	Remark
Address:-		Dhanora			1	pH	7.9	General
					2	Electrical Conductivity	0.2	General
Taluka:-		Wardha			3	Organic Carbon %(C)	0.62	iciently-enough
District:-		Wardha			4	Nitrogen (N)	434	iciently-enough
Survey No:-		109			5	Phosphorous (P)	23.63	less
					6	Pottassium (K)	240.6	iciently-enough

Sr. No.	Crop	Compost Manure (Ton per Hector)	Total quantity of fertilizers			Details	Fertilizers to be given to crops		
			Nitrogen (N)	Phosphorus (P)	Pottassium (K)		Urea	Single Super Phosphate	Murate of Potash
1	Cotton		50	25	25	At the time of sowing	49	195	38
		7	45	31	22.5	After 30 days	49		
2	Soyabeen	5	30 / 27	75 / 94		At the time of sowing	80	586	
3	Jowar	6	80	40	40	At the time of sowing	78	313	60
			72	50	36	After 30 days	78		
4	Graundnut	5	25 / 22.5	50 / 62.5		At the time of sowing	49	391	
5	Pigeon	5	25 / 22.5	50 / 62.5		At the time of sowing	49	391	

Farmer's Name :-		Jayashree Rameshrao Kahate			Sr. No.	General Analysis of soil	Observation	Remark
Address:-		Rohana			1	pH	8.06	General
					2	Electrical Conductivity	0.14	General
Taluka:-		Arvi			3	Organic Carbon %(C)	0.35	less
District:-		Wardha			4	Nitrogen (N)	245	less
Survey No:-					5	Phosphorous (P)	17.78	less
					6	Pottassium (K)	239.2	Medium

Sr. No.	Crop	Compost Manure (Ton per Hector)	Total quantity of fertilizers			Details	Fertilizers to be given to crops		
			Nitrogen (N)	Phosphorus (P)	Pottassium (K)		Urea	Single Super Phosphate	Murate of Potash
1	Cotton		50	25	25	At the time of sowing	68	195	42
		7	62.5	31	25	After 30 days	68		
2	Soyabeen	5	30 / 37.5	75 / 94		At the time of sowing	81.5	586	
3	Jowar	6	80	40	40	At the time of sowing	55	313	67
			100	50	40	After 30 days	55		
4	Graundnut	5	25 / 31	50 / 62.5		At the time of sowing	67	391	
5	Pigeon	5	25 / 31	50 / 62.5		At the time of sowing	67	391	

Farmer's Name :- Taibai Jamnare
Address:- Rohana
Taluka:- Arvi
District:- Wardha
Survey No:-

Sr. No.	General Analysis of soil	Observation	Remark
1	pH	8.07	General
2	Electrical Conductivity	0.21	General
3	Organic Carbon %(C)	0.54	Medium
4	Nitrogen (N)	378	Medium
5	Phosphorous (P)	18.9	Less
6	Pottassium (K)	240.6	iciently-eno

Sr. No.	Crop	Compost Manure (Ton per Hector)	Total quantity of fertilizers			Details	Fertilizers to be given to crops		
			Nitrogen (N)	Phosphorous (P)	Pottassium (K)		Urea	Single Super Phosphate	Murate of Potash
1	Cotton	7	50	25	25	At the time of sowing	55	195	38
			50	31	22.5	After 30 days	55		
2	Soyabeen	5	30 / 45	75 / 94		At the time of sowing	65	586	
3	Jowar	6	80	40	40	At the time of sowing	87	313	60
			80	50	36	After 30 days	87		
4	Graundnut	5	25 / 25	50 / 62.5		At the time of sowing	54	391	
5	Pigeon	5	25 / 25	50 / 62.5		At the time of sowing	54	391	

Farmer's Name :- Gajanan Patekar
Address:- Rohana
Taluka:- Arvi
District:- Wardha
Survey No:- 219

Sr. No.	General Analysis of soil	Observation	Remark
1	pH	8.02	General
2	Electrical Conductivity	0.23	General
3	Organic Carbon %(C)	0.27	less
4	Nitrogen (N)	189	less
5	Phosphorous (P)	15.3	less
6	Pottassium (K)	247.3	iciently-eno

Sr. No.	Crop	Compost Manure (Ton per Hector)	Total quantity of fertilizers			Details	Fertilizers to be given to crops		
			Nitrogen (N)	Phosphorous (P)	Pottassium (K)		Urea	Single Super Phosphate	Murate of Potash
1	Cotton	7	50	25	25	At the time of sowing	68	195	38
			62.5	31	22.5	After 30 days	68		
2	Soyabeen	5	30 / 37.5	75 / 94		At the time of sowing	81.5	586	
3	Jowar	6	80	40	40	At the time of sowing	55	313	60
			100	50	36	After 30 days	55		
4	Graundnut	5	25 / 31	50 / 62.5		At the time of sowing	67	391	
5	Pigeon	5	25 / 31	50 / 62.5		At the time of sowing	67	391	

Study of Physico-Chemical Properties...

Farmer's Name :-	Jamnalal Bhagat				Sr. No.	General Analysis of soil	Observation	Remark
Address:-	Borkha				1	pH	7.8	General
					2	Electrical Conductivity	0.23	General
Taluka:-	Samudrapur				3	Organic Carbon %(C)	0.59	Medium
District:-	Wardha				4	Nitrogen (N)	413	Medium
Survey No:-	14/4				5	Phosphorous (P)	25.88	less
					6	Pottassium (K)	309.1	enough

Sr. No.	Crop	Compost Manure (Ton per Hector)	Total quantity of fertilizers			Details	Fertilizers to be given to crops		
			Nitrogen (N)	Phosphorous (P)	Pottassium (K)		Urea	Single Super Phosphate	Murate of Potash
1	Cotton		50	25	25	At the time of sowing	55	195	32
		7	50	31	19	After 30 days	55		
2	Soyabeen	5	30 / 30	75 / 94		At the time of sowing	65	586	
3	Jowar	6	80	40	40	At the time of sowing	87	313	50
			80	50	30	After 30 days	87		
4	Graundnut	5	25 / 25	50 / 62.5		At the time of sowing	54	391	
5	Pigeon	5	25 / 25	50 / 62.5		At the time of sowing	54	391	

Farmer's Name :-	Sukhdeorao Tiwade				Sr. No.	General Analysis of soil	Observation	Remark
Address:-	Chikhalkot				1	pH	7.64	General
					2	Electrical Conductivity	0.2	General
Taluka:-	Samudrapur				3	Organic Carbon %(C)	0.64	iciently-enough
District:-	Wardha				4	Nitrogen (N)	448	iciently-enough
Survey No:-	18/1 B				5	Phosphorous (P)	21.38	less
					6	Pottassium (K)	346.8	Enough

Sr. No.	Crop	Compost Manure (Ton per Hector)	Total quantity of fertilizers			Details	Fertilizers to be given to crops		
			Nitrogen (N)	Phosphorous (P)	Pottassium (K)		Urea	Single Super Phosphate	Murate of Potash
1	Cotton		50	25	25	At the time of sowing	49	194	32
		7	45	31	19	After 30 days	49		
2	Soyabeen	5	30 / 27	75 / 94		At the time of sowing	59	588	
3	Jowar	6	80	40	40	At the time of sowing	78	313	50
			72	50	30	After 30 days	78		
4	Graundnut	5	25 / 22.5	50 / 62.5		At the time of sowing	49	391	
5	Pigeon	5	25 / 22.5	50 / 62.5		At the time of sowing	49	391	

Farmer's Name :-	Vasanta Khandale			Sr. No.	General Analysis of soil	Observation	Remark
Address:-	Kelzar			1	pH	7.81	General
				2	Electrical Conductivity	0.19	General
Taluka:-	Seloo			3	Organic Carbon %(C)	0.5	Medium
District:-	Wardha			4	Nitrogen (N)	350	Medium
Survey No:-	522			5	Phosphorous (P)	9.9	Very less
				6	Pottassium (K)	239.2	Medium

Sr. No.	Crop	Compost Manure (Ton per Hector)	Total quantity of fertilizers			Details	Fertilizers to be given to crops		
			Nitrogen (N)	Phosphorous (P)	Pottassium (K)		Urea	Single Super Phosphate	Murate of Potash
1	Cotton		50	25	25	At the time of sowing	55	238	42
		7	50	38	25	After 30 days	55		
2	Soyabeen	5	30 / 30	75 / 113		At the time of sowing	65	706	
3	Jowar	6	80	40	40	At the time of sowing	87	375	69
			80	60	40	After 30 days	87		
4	Graundnut	5	25 / 25	50 / 75		At the time of sowing	54	469	
5	Pigeon	5	25 / 25	50 / 75		At the time of sowing	54	469	

Farmer's Name :-	Devidas Katarkar			Sr. No.	General Analysis of soil	Observation	Remark
Address:-	Rasulabad			1	pH	7.82	General
				2	Electrical Conductivity	0.2	General
Taluka:-	Arvi			3	Organic Carbon %(C)	0.82	enough
District:-	Wardha			4	Nitrogen (N)	574	enough
Survey No:-				5	Phosphorous (P)	15.75	Less
				6	Pottassium (K)	185.5	Medium

Sr. No.	Crop	Compost Manure (Ton per Hector)	Total quantity of fertilizers			Details	Fertilizers to be given to crops		
			Nitrogen (N)	Phosphorous (P)	Pottassium (K)		Urea	Single Super Phosphate	Murate of Potash
1	Cotton		50	25	25	At the time of sowing	41	194	42
		7	38	31	25	After 30 days	41		
2	Soyabeen	5	30 / 22.5	75 / 94		At the time of sowing	49	588	
3	Jowar	6	80	40	40	At the time of sowing	65	313	69
			60	50	40	After 30 days	65		
4	Graundnut	5	25 / 19	50 / 62.5		At the time of sowing	41	391	
5	Pigeon	5	25 / 19	50 / 62.5		At the time of sowing	41	391	

Farmer's Name :-	Devidas Namawre			Sr. No.	General Analysis of soil	Observation	Remark
Address:-	Vijaygopal			1	pH	7.66	General
				2	Electrical Conductivity	0.4	General
Taluka:-	Deoli			3	Organic Carbon %(C)	0.51	Medium
District:-	Wardha			4	Nitrogen (N)	357	Medium
Survey No:-	521/53			5	Phosphorous (P)	8.77	Very less
				6	Pottassium (K)	284	iciently-enough

Sr. No.	Crop	Compost Manure (Ton per Hector)	Total quantity of fertilizers			Details	Fertilizers to be given to crops		
			Nitrogen (N)	Phosphorous (P)	Pottassium (K)		Urea	Single Super Phosphate	Murate of Potash
1	Cotton		50	25	25	At the time of sowing	55	238	38
		7	50	37.5	22.5	After 30 days	55		
2	Soyabeen	5	30 / 30	75 / 113		At the time of sowing	65	706	
3	Jowar	6	80	40	40	At the time of sowing	87	375	60
			80	60	36	After 30 days	87		
4	Graundnut	5	25 / 25	50 / 75		At the time of sowing	54	469	
5	Pigeon	5	25 / 25	50 / 75		At the time of sowing	54	469	

Farmer's Name :-	Anandrao Tegade			Sr. No.	General Analysis of soil	Observation	Remark
Address:-	Shirpur			1	pH	7.62	General
				2	Electrical Conductivity	0.19	General
Taluka:-	Deoli			3	Organic Carbon %(C)	0.46	Medium
District:-	Wardha			4	Nitrogen (N)	322	Medium
Survey No:-	240			5	Phosphorous (P)	13.73	Very Less
				6	Pottassium (K)	255.4	iciently-eno

Sr. No.	Crop	Compost Manure (Ton per Hector)	Total quantity of fertilizers			Details	Fertilizers to be given to crops		
			Nitrogen (N)	Phosphorous (P)	Pottassium (K)		Urea	Single Super Phosphate	Murate of Potash
1	Cotton		50	25	25	At the time of sowing	55	238	38
		7	50	37.5	22.5	After 30 days	55		
2	Soyabeen	5	30 / 30	75 / 113		At the time of sowing	65	706	
3	Jowar	6	80	40	40	At the time of sowing	87	375	60
			80	60	36	After 30 days	87		
4	Graundnut	5	25 / 25	50 / 75		At the time of sowing	54	469	
5	Pigeon	5	25 / 25	50 / 75		At the time of sowing	54	469	

Farmer's Name :-	Dayaram Kadve			Sr. No.	General Analysis of so	Observatio	Remark
Address:-	Bhalewadi			1	pH	7.9	General
				2	Electrical Conductivity	0.21	General
Taluka:-	Karanja			3	Organic Carbon %(C)	0.21	Less
District:-	Wardha			4	Nitrogen (N)	147	Less
Survey No:-6/2	240			5	Phosphorous (P)	16.65	less
				6	Pottassium (K)	258	ficintly-enou

Sr. No.	Crop	Compost Manure (Ton per Hector)	Total quantity of fertilizers			Details	Fertilizers to be given to crops		
			Nitrogen (N)	Phosphorous (P)	Pottassium (K)		Urea	Single Super Phosphate	Murate of Potash
1	Cotton		50	25	25	At the time of sowing	68	195	38
		7	62.5	31	22.5	After 30 days	68		
2	Soyabeen	5	30 / 38	75 / 94		At the time of sowing	81	588	
3	Jowar	6	80	40	40	At the time of sowing	55	313	60
			100	50	36	After 30 days	455		
4	Graundnut	5	25 / 31.25	50 / 63		At the time of sowing	68	394	
5	Pigeon	5	25 / 31.25	50 / 63		At the time of sowing	68	394	

Farmer's Name :-	Mahadeo Patil			Sr. No.	General Analysis of so	Observatio	Remark
Address:-	Jaurwada			1	pH	7.8	General
				2	Electrical Conductivity	0.2	General
Taluka:-	karanja			3	Organic Carbon %(C)	0.3	less
District:-	Wardha			4	Nitrogen (N)	210	less
Survey No:-	52			5	Phosphorous (P)	11.03	Very less
				6	Pottassium (K)	252	iciently-enou

Sr. No.	Crop	Compost Manure (Ton per Hector)	Total quantity of fertilizers			Details	Fertilizers to be given to crops		
			Nitrogen (N)	Phosphorous (P)	Pottassium (K)		Urea	Single Super Phosphate	Murate of Potash
1	Cotton		50	25	25	At the time of sowing	68	238	38
		7	63	37.5	22.5	After 30 days	68		
2	Soyabeen	5	30 / 38	75 / 113		At the time of sowing	82	706	
3	Jowar	6	80	40	40	At the time of sowing	55	375	40
			120	60	36	After 30 days	55		
4	Graundnut	5	25 / 31	50 / 75		At the time of sowing	67	469	
5	Pigeon	5	25 / 31	50 / 75		At the time of sowing	67	469	

Farmer's Name :-	Ashok Dhole			Sr. No.	General Analysis of soil	Observation	Remark
Address:-	Jasapur			1	pH	8	General
				2	Electrical Conductivity	0.27	General
Taluka:-	karanja			3	Organic Carbon %(C)	0.28	Less
District:-	Wardha			4	Nitrogen (N)	196	Less
Survey No:-	97			5	Phosphorous (P)	11.7	very less
				6	Pottassium (K)	264.8	iciently-eno

Sr. No.	Crop	Compost Manure (Ton per Hector)	Total quantity of fertilizers			Details	Fertilizers to be given to crops		
			Nitrogen (N)	Phosphorous (P)	Pottassium (K)		Urea	Single Super Phosphate	Murate of Potash
1	Cotton		50	25	25	At the time of sowing	68	238	38
		7	62.5	37.5	22.5	After 30 days	68		
2	Soyabeen	5	30 / 38	75 / 113		At the time of sowing	81.5	706	
3	Jowar	6	80	40	40	At the time of sowing	55	375	40
			100	60	36	After 30 days	55		
4	Graundnut	5	25 / 31	50 / 75		At the time of sowing	67	469	
5	Pigeon	5	25 / 31	50 / 75		At the time of sowing	67	469	

Farmer's Name :-	Ajij Khan Pathan			Sr. No.	General Analysis of soil	Observation	Remark
Address:-	Wadi			1	pH	8.24	General
				2	Electrical Conductivity	0.22	General
Taluka:-	Ashti			3	Organic Carbon %(C)	0.32	Less
District:-	Wardha			4	Nitrogen (N)	224	Less
Survey No:-	407			5	Phosphorous (P)	15.75	Less
				6	Pottassium (K)	231	Medium

Sr. No.	Crop	Compost Manure (Ton per Hector)	Total quantity of fertilizers			Details	Fertilizers to be given to crops		
			Nitrogen (N)	Phosphorous (P)	Pottassium (K)		Urea	Single Super Phosphate	Murate of Potash
1	Cotton		50	25	25	At the time of sowing	68	194	42
		7	62.5	31	25	After 30 days	68		
2	Soyabeen	5	30 / 38	75 / 94		At the time of sowing	81.5	588	
3	Jowar	6	80	40	40	At the time of sowing	55	313	67
			100	50	40	After 30 days	55		
4	Graundnut	5	25 / 31	50 / 62.5		At the time of sowing	67	391	
5	Pigeon	5	25 / 31	50 / 62.5		At the time of sowing	67	391	

Farmer's Name :-	Marotrao Tekam					Sr. No.	General Analysis of soil	Observation	Remark
Address:-	Yar					1	pH	8.23	General
						2	Electrical Conductivity	0.23	General
Taluka:-	Ashti					3	Organic Carbon %(C)	0.3	less
District:-	Wardha					4	Nitrogen (N)	210	less
Survey No:-						5	Phosphorous (P)	13.95	Very less
						6	Pottassium (K)	193.5	Medium

Sr. No.	Crop	Compost Manure (Ton per Hector)	Total quantity of fertilizers			Details	Fertilizers to be given to crops		
			Nitrogen (N)	Phosphorous (P)	Pottassium (K)		Urea	Single Super Phosphate	Murate of Potash
1	Cotton		50	25	25	At the time of sowing	68	238	42
		7	62.5	38	25	After 30 days	68		
2	Soyabeen	5	30 / 38	75 / 113		At the time of sowing	81.5	706	
3	Jowar	6	80	40	40	At the time of sowing	55	375	67
			100	60	40	After 30 days	55		
4	Graundnut	5	25 / 31	50 / 75		At the time of sowing	67	469	
5	Pigeon	5	25 / 31	50 / 75		At the time of sowing	67	469	

Farmer's Name :-	Anjani Gayakwad					Sr. No.	General Analysis of soil	Observation	Remark
Address:-	Tigaon					1	pH	7.62	General
						2	Electrical Conductivity	0.11	General
Taluka:-	Ashti					3	Organic Carbon %(C)	0.16	Very less
District:-	Wardha					4	Nitrogen (N)	112	Very less
Survey No:-	227					5	Phosphorous (P)	15.53	less
						6	Pottassium (K)	205.6	Medium

Sr. No.	Crop	Compost Manure (Ton per Hector)	Total quantity of fertilizers			Details	Fertilizers to be given to crops		
			Nitrogen (N)	Phosphorous (P)	Pottassium (K)		Urea	Single Super Phosphate	Murate of Potash
1	Cotton		50	25	25	At the time of sowing	82	195	42
		7	75	31	25	After 30 days	82		
2	Soyabeen	5	30 / 45	75 / 94		At the time of sowing	98	586	
3	Jowar	6	80	40	40	At the time of sowing	132	313	67
			120	50	40	After 30 days	132		
4	Graundnut	5	25 / 37.5	50 / 62.5		At the time of sowing	82	391	
5	Pigeon	5	25 / 37.5	50 / 62.5		At the time of sowing	82	391	

| CHAPTER VI | SUMMARY AND CONCLUSIONS |

For increasing the soil fertility and crop yields the vast studies have been done India. As a result of these studies, suitable analytical methods have been recommended and fertilizer recommendation prescribed.

In order that fertilizer recommendations become more useful, more calibration studies of soil test with responses of different crops should be undertaken under all agro-climatic conditions in country. Accurate fertilizer recommendation based on soil-test-crop-response correlation studies in countries will make the soil testing service more useful and popular and will result in the more efficient use of fertilizers to obtain crop yields.

The district Wardha is surrounded by small towns and villages. The main income source of the people is agriculture. The main crops yielded in this region are cotton, jowar, soyabean, sugarcane, bananas, pulses and some common vegetables.

To increase the crop growth and yield of crops, the farmers used common fertilizers containing N, KandP elements and superphosphates etc. But, for the growth of plants as mentioned earlier, micronutrients are also essential.

About 35 soil samples from different villages were collected and physical parameters (pH, electrical conductivity, etc.) as well as available nutrients (carbon, available nitrogen, phosphorus, and potassium) present in the soil, taken from various parts of this region were detected. Available micronutrients are also determined by known methods. After studying all these properties, the suitable fertilizer recommendation was prescribed so that quality of soil can be improved and thus crops productivity can also be increased.

In general it is observed that

1. The soil in Wardha district is slightly alkaline (pH 7.5-8.5).

2. The electrical conductivity of soil varies from place to place (0.20-0.40).

3. The available carbon is in 'very less' to 'medium' range.

4. The available nitrogen is also in 'very less' to 'medium' range.

5. Available phosphorous in about 80% of soil in Wardha region is very less; while in 20% soil it is medium.

6. The available potassium is found to be sufficiently enough.

7. There is a deficiency of zinc while generally other micronutrients (copper, iron, manganese) are found to be sufficient.

Following fertilizers are recommended to the farmers as per requirement.

1. Urea

2. Muriate of potash

3. Single super phosphate

4. Zinc sulphate (25 to 30 Kg./Hector directly or 0.4% by spraying.)

In addition to interrelationship among soil, crop and fertilizer, crop rotation and management is also considered. The fertilizer applications also supplemented with the additions of farm manures,

crop residues and green manures, especially under scarcity conditions of fertilizers and their high cost. In summary, it may be reiterated that fertilizer practices are an important phase of the fertility management of soils in crop production.

REFERENCES

> Soil and the environment – Alan Wild, Cambridge University press

> The nature and properties of soils – Nyle C. Brady, Euracia publishing house

> Land and soil – Dr. S.P.Raychaudhary,National book trust ,India.

> Soil science simplified – Milo I. Harpstead and Francis D.Hole, Scientific publisher, Jodhpur.

> The ABC of soils – Jacob S. Joffe, Oxford Book Company

> Soils in our environment – R.W. Miler and R.L.Donahure, Prentice hall of India.

> Soil Microbiology – M.S.Coyne, Delmar- Thomson learning publication

> Soils and soil fertility- Lauis M.Thompson and Frederic R. Troeh

➢ Introduction to the principles and practice of soil science – R.E. White, Blackwell Scientific publications Ltd.

➢ Chemistry of the soil – F.E.Bear, Oxford and IBH publishing company

➢ A Handbook of soil analysis – C.Horuld Wright, Logus press

➢ Introduction to agriculture geology soil physics and agriculture climatology –Dr. K.S. Yawalkar

➢ A text-book of soil physics – M.C. Oswal, Vikas publishing house, New Delhi.

➢ Soil physics-L.D.Baver, walterH.Gardner and W.R.Gardner, Wiley eastern Ltd.,
New Delhi

    ➢ Analytical agriculture chemistry – Kanwar J.S. and Chopra S.L.

➢ Nature and property of soil – Nycle C. Brady.

➢ Principles of soil science – M.M.Rai, Macmillan publication

➢ Textbook of soil science – T.D. Biswas and S.K. Mukherjee, Tata Mcgraw-Hill Publishing Company Limited, New Delhi

➢ Soil fertility and fertilizers – Samuel L. Tisdale, Prentice Hall of India Private Limited, New Delhi

➢ Methods in environmental analysis water, soil and air – P.K. Gupta, Agrobios (India)

➢ Introduction to agronomy and soil and water management – V.G. Vaidya and K.R. Sahastrabuddhe – Continental Prakashan, Pune

➢ A Practical manual: Methods of soil, water and plant analysis – Dr. M.N. Patil and Dr. S.K. Thakre

➢ Baethgan, W.E. and Alley, M.M.1989. A manual colorimetric method of measuring ammonium N in soil and plant Kjeldahl digest. Comm. Soil Sci. Plant Anal.20:961-969.

➢ Bray,R.H. and Kurtg, L.T. 1945. Determination of total, organic and available forms of phosphorous in the soil.

➤ Carter, M.R. 1993. Soil sampling and method of analysis. Ed. Canadian Soc. Soil Sci. Lewis publishers, USA, 823

➤ Charles, M.J. and Simmons, M.S.1986. Methods for the determination of carbon in soils and sediments. A review. Analyst 111:385-390.

➤ Gardner, W.H. 1986. Water content. In A. Klute, Ed. Methods of soil analysis. Part 1. Agronomy No.9. American Societyof Agronomy, Madison, WI. 493-544.

➤ Ghosh, A.B., Bajaj, J.C., Hasan, R and Singh, D. 1983. Soil and water testing methods, A laboratory manual, IARI, New Delhi. India.

➤ Grava, J.1980. Importance of soil extraction techniques in recommended soil test procedures for the north central region. Bull. 499 North Dakota Agric. Exp. Stn., North Dakota State University, Fargo.9-11.

➤ Hesse, P.R. 1971. A Text book of soil chemical analysis. John Murry Ltd. London, UK: 528.

➢ Jackson, M.L. 1962. Soil chemical analysis, constable and Co. Ltd., London.

➢ Jackson, M.L., 1973.Soil Chemical analysis, Prentice Hall of India Pvt. Ltd. New Delhi.

➢ Jones, J.B., jr.1985. Soil testing and plant analysis: guides to the fertilization of horticulture crops. Hortic. Rev.7:1-68.

➢ Kalra, Y.P. and Maynard, D.G. 1991. Methods manual for forest soil and plant anallysis. For. can. Edmonton, Alberta, Canada: 116.

➢ Knudsen. D., Paterson, G.A., and pratt, P.F.1982. Lithium, Sodium and Potassium. Pages 225-246 in A.L. Page et al., eds. Methods of soil analysis. Part 2. 2nd ed. Agronomy society of agronomy, Madison, WI.

➢ Lindsay, W.L. and Norvell, W.A.1978, development of DTPA soil test for zinc, manganese and copper. Soil Sci, Soc.Am.J.42:421-428.

➢ Nelson, D.W. and Sommaers, L.E.1982. Total carbon, organic carbon and organic matter. In methods of soil analysis, part 2 (Eds Page et al.) Argon.1 Am. Soc. Argon., Madison, Wisconsin; 539-579.

➢ Palaskar, M.S.Babrekar,P.G. and Ghosh, A.B.1981. A rapid analytical technique to estimate sulpher in soil and plant extracts. J. Indian Soc. Soil Sci. 29:249-256.

ANNEXURE- I

S.N.	INFORMATION	
1	Name of the cultivator	
2	Address	
3	Field No./Local name of the field	
4	Slope	
5	Drainage	
6	Irrigation	
7	Previous cropping history	
8	Fertilizer used for previous crops	
9	Fertilizer used for proposed crops	
10	Date of sampling	
11	Depth of sampling	

Printed in Great Britain
by Amazon